はしがき

天災・人災、被害者の身元確認情報に効果を発揮するのが、この靴であり、この靴の構成・用途・使用方法などを独創的に表現する。例えば、歯型などで身元を確認する方法は周知の通りである。しかし、靴で身元を確認する方法の表現は例が見られないので、これをイラスト解説で著作権の表現をする。

Preface

These shoes demonstrate an effect to a natural disaster and a man-made disaster, and a victim's body identification information, and it expresses creatively the composition, the use, the directions, etc. for these shoes.

For example, the method of checking an identity with a denture mold etc. is well known.

However, since an example is not seen, expression of a method which checks an identity with shoes expresses copyright for this by illustration description.

目　次

1、天災・人災　被害者の身元確認情報靴　情報携帯靴　ワカール（イラスト解説）

　(1)　紳士靴 -- 5

　(2)　女性靴 -- 6

　(3)　運動靴 -- 7

　(4)　長靴 -- 8

　(5)　作業靴 -- 9

　(6)　登山靴 --- 10

　(7)　デッキシューズ --- 11

　(8)　スキー靴 --- 12

2、English of the usage

A natural disaster and man-made disaster　　Victim body identification information shoes　Portable information shoes　Ｗａｋａｒｌ

(Illustration explanation)

(1) -- 13

Gentleman shoe

(2) -- 14

Female shoes

(3) -- 15

Sports shoes

(4) -- 16

Boots

(5)--17

Work shoes

(6)--18

Mountain-climbing boots

(7)--19

Deck shoes

(8)--20

Ski boots

3、公報解説--21

4、Patent journal English --33

1、天災・人災 被害者の身元確認情報靴 情報携帯靴 ワカール（イラスト解説）

　靴の中に個人情報カードをセットで身元確認

(1) 紳士靴

(2) 女性靴

(3) 運動靴

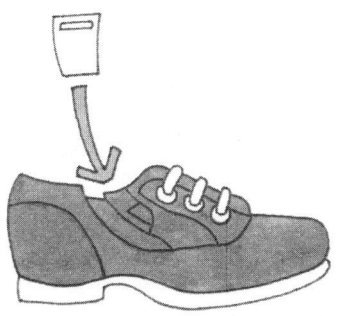

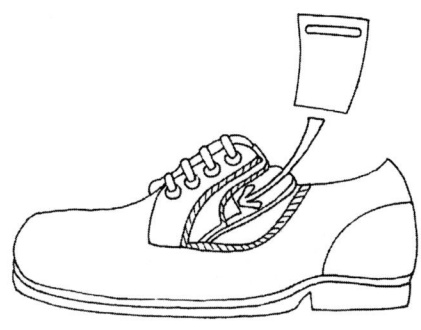

(4) 長靴

(5) 作業靴

生き埋め・土砂崩れ

(6) 登山靴

遭難

(7) デッキシューズ

転覆

(8) スキー靴

雪崩

2、English of the usage

A natural disaster and man-made disaster　Victim body identification information shoes　Portable information shoes　Ｗａｋａｒｌ

(Illustration explanation)

The body of a personal information card is identified by the set in shoes.

(1) Gentleman shoe

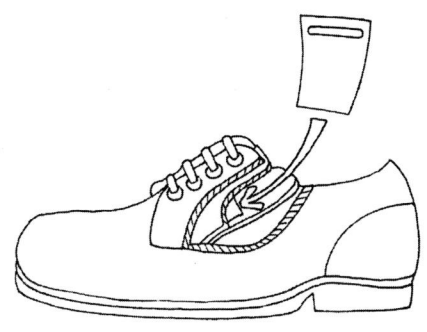

(2) Female shoes

(3) Sports shoes

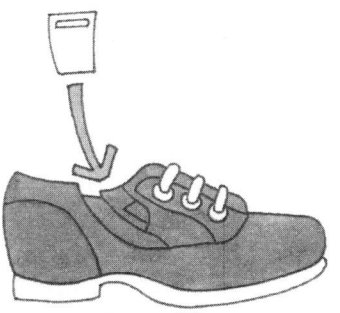

(4) Boots

(5) Work shoes

Mudslide

(6) Mountain-climbing boots

disaster

(7) Deck shoes

　　Overthrow

(8) Ski boots

Snowslide

3、公報解説

実用新案登録第３１７８５３２号

考案の名称；情報携帯機能を備えた靴の構造

実用新案権者；斉藤　通

【要約】　　　（修正有）

【課題】屋外で事故などに被災した場合に、確実に本人の確認ができる情報携帯機能を備えた靴を提供する。

【解決手段】靴のアッパー１の裏面であって、舌革１１や爪先皮１２、先芯１３の裏面などに、耐水性の、防水チャック３１で挿入口を構成したカード収納容器３を設けて構成する。靴の一部に足首締結帯４を取りつける。カード収納容器３には、身分を証明する事項を記入したカードなどを挿入して使用する。又、足首締結帯４を足首に結び、非常時に靴が離脱しても近辺に保持できる特徴がある。

【選択図】図１

【実用新案登録請求の範囲】

【請求項１】

靴を底板とアッパーとで区分した場合に、

そのアッパーの裏面であって、舌革や爪先皮、先芯などの裏面に、

防水チャックで挿入口を構成した耐水性のカード収納容器を設け、

靴の一部に足首締結帯を取り付けて構成した、

情報携帯機能を備えた靴。

【請求項２】

靴を底板とアッパーとで区分した場合に、そのアッパーの表面の前面、側面、背面

に、防水チャックで挿入口を構成した耐水性のカード収納容器を設け、靴の一部に足首締結帯を取り付けて構成した、情報携帯機能を備えた靴。

【考案の詳細な説明】

【技術分野】

【0001】

本考案は、氏名、住所、血液型などの情報を常時携帯することができ、事故の場合でも確実に本人の確認ができる、情報携帯機能を備えた靴の構造に関するものである。

【背景技術】

【0002】

屋外で事故が発生した場合に、その被害者の個人の特定が必要であるが、その特定が困難な場合が実際に発生している。

特に地震や津波のように大勢の負傷者、死者が同時に発生した場合には、その身元の確認、特定がきわめて困難となっている。

そのために、個人情報を記載したカードなどを常時ポケットに入れて携帯していれば確認は容易であるはずである。

【考案の概要】

【考案が解決しようとする課題】

【0003】

しかし一般の人は、季節や好み、目的に応じて相当多種類の衣服を所持している。それらの人が、すべての衣服に情報を記入したカードを備えること、それを常に持ち歩くことはほとんど不可能である。

さらに衣服は頻繁に洗濯をするから、その度にカードを取り出して、洗濯後に収納

るという行為もきわめて煩わしく実際的ではない。

【課題を解決するための手段】

【０００４】

上記のような従来の問題を改善するためになされた本考案の情報携帯機能を備えた靴の構造は、靴のアッパーの裏面であって、舌革や爪先皮、先芯の裏面に耐水性の、防水チャックで挿入口を構成したカード収納容器を設け靴の一部には足首締結帯を取り付けて構成したものである。

【考案の効果】

【０００５】

本考案の情報携帯機能を備えた靴の構造は上記したようになるから、以下のような効果の少なくともひとつを得ることができる。

＜１＞本考案の構成の靴を購入した購入者は、蓋付きの容器に身分を証明する事項を記入したカードなどを挿入して使用する。

＜２＞そして外出時には、必ず靴を履いており、しかも靴はひもやゴムなどで脱げにくく構成してあるから、被害者の身体から離れにくい。

＜３＞この点で、下着や上着、コートのポケットに身分情報を備えていれば同じ効果を得られるはずであるが、上記したように実際には衣服は多数の点数を所持しており、洗濯の回数も多いので、常にカードを携帯することは困難である。

＜４＞それに対して一般人では靴の所有数はそれほど多くはなく、被服のように季節や好み、目的に応じて多数、多種類の靴を持っている人は少なく、洗濯の機会も少ない。

＜５＞よって靴に蓋つきの収納容器が設けてあれば、数点の靴に情報を保存しておくだけで、身分情報を確保することができる。

＜６＞さらに本考案の情報携帯機能を備えた靴は、靴の一部に足首締結帯を備えている。そのために例えば「洗濯機に巻き込まれたような衝撃」と表現されるような津波に巻き込まれて靴が足から脱げたとしても、靴は足首から離脱することがなく、確実に情報を保持することができる。

＜７＞児童や学生、老人などは運転免許証のような身分証明書を持っていない場合が多いが、屋外で靴を履いていない場合はめったにない。したがって本考案の情報携帯機能を備えた靴を履いていれば万一事故に会った場合でも個人の情報を常に携帯することができる。

【図面の簡単な説明】

【０００６】

【図１】革靴における実施例の説明図。

【図２】運動靴における実施例の説明図。

【図３】靴が脱げた場合の説明図。

【実施例】

【０００７】

次に本考案の情報携帯機能を備えた靴の構造の実施例について説明する。

【０００８】

＜１＞靴の構造

まず一般の靴の構造について説明すると、靴はアッパー１と底２で区分することができる。

すなわちアッパー１とは底２を除いた上の部分の総称である。

そして本考案の対象とする靴は、革靴、運動靴、ハイヒールなどすべての靴を対象とすることができる。

【０００９】

＜２＞収納容器

収納容器３とは防水材料で構成した袋体である。

その一部には防水チャック３１のような開閉自在の挿入口を開口する。

その防水チャック３１部分から、個人情報を記入したカードを挿入することができる。

個人情報とは靴を履いている本人の氏名、住所、電話、血液型、既往症、親族の氏名、学校や会社名などである。

【００１０】

＜３＞取り付け場所

収納容器３は、靴のアッパー１の一部の裏側に接着、縫込みなどで設置する。

例えば革靴の場合には、舌革１１と称する部分の裏面、あるいは爪先皮１２の裏面、あるいは先芯
１３の裏面、踵の裏面などである。

運動靴などでは、デザイン的に美観を損なわなければ、アッパー１の表面、すなわちアッパー１の上面、側面、背面に、前記の収納容器を取り付けることもできる。

その場合にも収納容器３の挿入口は防水チャック３１で保護しておけば、個人情報のカードが汚れたり破損することがない。

【００１１】

＜４＞足首締結帯

上記の情報携帯機能を備えた靴の構造の一部には、足首締結帯４を取りつける。

足首締結帯４は通常の繊維、合成繊維、伸縮する材料、チェーンなどの長い帯体、紐体で構成する。

そのように、帯状の材料だけに限らず、紐状の材料も含めて本発明では足首締結帯４と称する。

【００１２】

＜４－１＞足首締結帯の取り付け足首締結帯４は、ヒールカップなどと称する踵側に、靴を履くときに引き上げるための円環状の「つまみ」１４が付いている場合には、そのつまみ１４を通して足首締結帯４を挿入して取り付ける。

運動靴では舌革の端部につまみ１４が取り付けてある構造のものも市販されているので、そのつまみ１４を通して足首締結帯４を挿入して取り付ける。

革靴など、つまみ１４のない靴の場合には、ゴムひもを通すような、短い筒状体を踵側に取り付け、その筒状体をつまみ１４として、そこに足首締結帯４を貫通させて取り付ける。

なお貫通させる場合だけではなく、一部を縫い付けるような点接触で取り付けることも可能である。

さらに図３に示すように、着用時にはつまみを貫通させた足首締結帯４を、つまみ１４の周囲でい

ったん結び、その後に足首締結帯４を足首に巻き付けることもできる。

【００１３】

＜４－２＞足首締結帯の長さ

足首締結帯４の長さは、靴を履いた場合にその足首を少なくとも１周し、前部あるいは後部で結ぶだけの余裕のある長さである。

足首締結帯４の一部にマジックテープ（登録商標）などと称する面接触体を取りつければ、あえて結ぶ必要はないから、それだけ短い寸法で足りる。

【００１４】

＜４－３＞ファッション性

この足首締結帯４を、色彩の豊かな帯や紐で構成しておけば、これを足首に巻き付けるときには、足元を飾るファッションとなる。

また足首に巻き付けない場合でも、靴の後ろのつまみ１４の周囲で花結びにしておけば、それだけでも装飾された靴として商品価値を高めることができる。

さらに多数の靴を店頭で並べて販売している場合に、本考案の靴だけはファッション性のある足首締結帯４を取り付けてあるから、購買者の目を引くことができ、類似商品との差別化を図り、売り上げに大きく貢献することができる。

【００１５】

＜４－４＞足首締結帯の表示効果

外出するときには当然靴を履くが、通常は数十足も保有している人は少なく、せいぜい数足程度である。

その程度の靴数であれば、すべての靴の収納容器に個人情報を記入したカードを挿入することもさほど困難ではない。

さらに運動靴以外では洗濯するということはほとんどないから、いったんカードを挿入しておけば

そのまま保持、保管することができる。

しかしたとえこの靴を履いていたとしても、万一事故にあった場合に警察や消防の担当者がその靴の内部の情報に気が付かなければ役に立たない。

その場合に本考案の靴であれば、つまみ１４か、あるいはいずれかの位置に足首締結帯４が取り付けてある。

したがって本考案の靴の機能、効果が広く知られれば、足首締結帯４が取りつけた

ある靴ならば個人情報が保管されている、と認識してもらうことができ、身元の確認を迅速に行うことができる。

【0016】

＜4-5＞靴が脱げた場合（図4）

前記した様に、津波の衝撃は「洗濯機に巻き込まれた」と表現されるほどの強烈なものであり、衣服などもはぎ取られる可能性がある。

また無謀運転の自動車事故に遭遇した場合にも、衝突の衝撃で靴が跳ね飛ばされる可能性もある。

したがって一般の靴であれば履いている靴がもぎ取られる可能性も十分に想定される。

しかし本考案の情報携帯機能を備えた靴の場合には足首締結帯4によって靴は足首にしっかりと取り付けてある。

そのために相当に大きな衝撃があっても、踵から爪先までの長さは、足首よりも大きいから靴が足から離れることがなく、本人情報を提示するという本来の機能を果たすことができる。

【符号の説明】

【0017】

1：アッパー

11：舌革

12：爪先皮

13：先芯

2：底

3：収納容器

３１：防水チャック

４：足首締結帯

【図面の簡単な説明】

【０００６】

【図１】革靴における実施例の説明図。

【図２】運動靴における実施例の説明図。

【図３】靴が脱げた場合の説明図。

図1

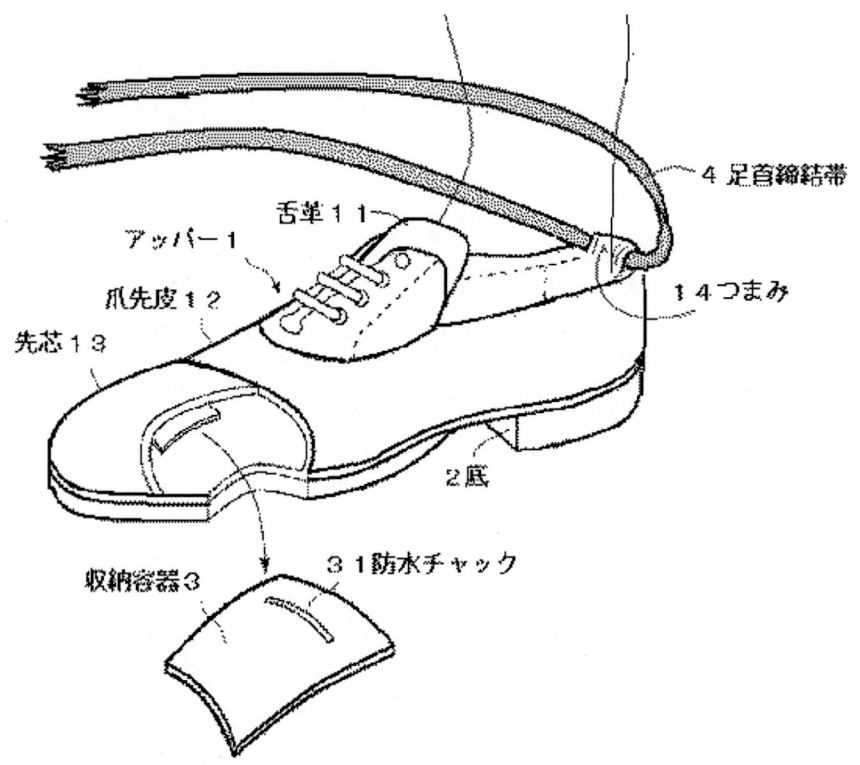

図2

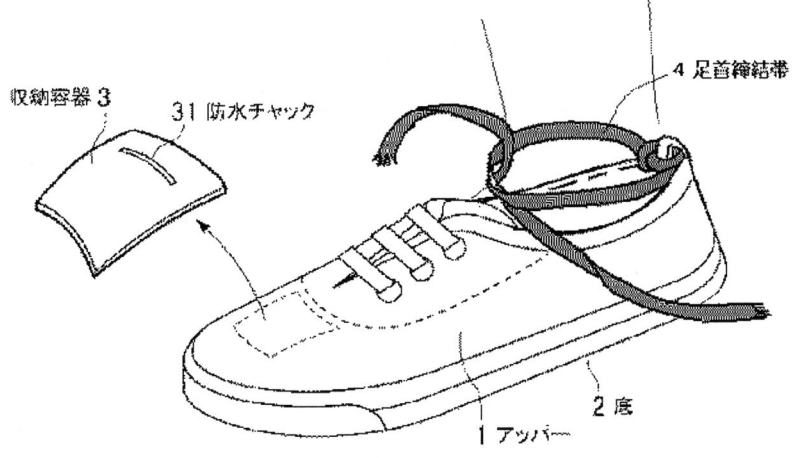

平常時に靴を履いている状態の説明図。
（靴の内部には個人情報カードが入っている）

図3

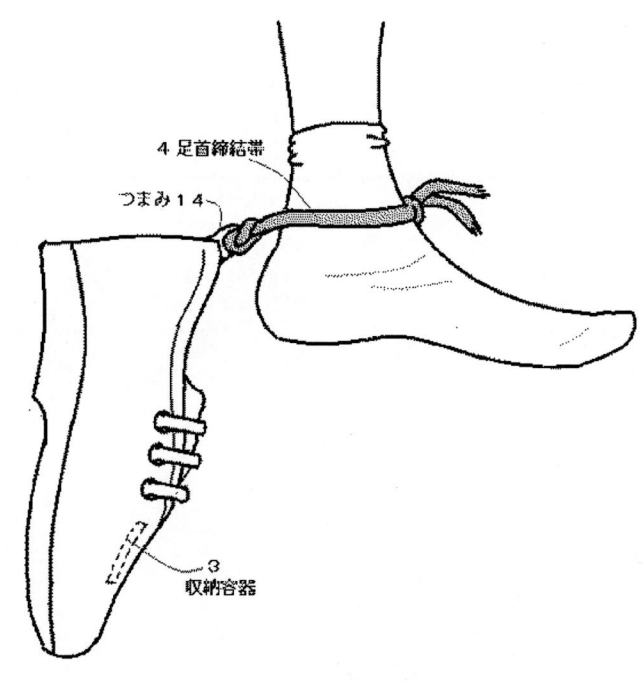

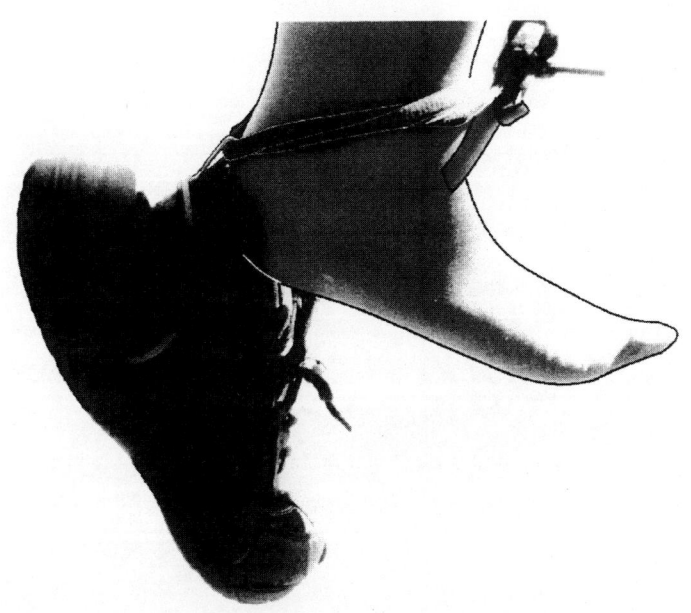

非常時に靴が脱げた状態の説明図。
（靴の内部の情報カードが、足から離れない）

4、Patent journal English

CLAIMS

[Claims]

[Claim 1]

When shoes are classified with a bottom plate and an upper,

It is a back surface of the upper and is at the back surface, such as a tongue, a tiptoe hide, a point core,

A waterproof card stowage container which constituted a loading slot from a water proof zipper is provided,

An ankle conclusion belt was attached and constituted in some shoes,

Shoes provided with an information portable function.

[Claim 2]

When shoes are classified with a bottom plate and an upper,

At a front face of the surface of the upper, a side surface, and a back face,

A waterproof card stowage container which constituted a loading slot from a water proof zipper is provided,

An ankle conclusion belt was attached and constituted in some shoes,

Shoes provided with an information portable function.

DETAILED DESCRIPTION

[Detailed explanation of the device]

[Field of the Invention]

[0001]

This design can always carry the information on a name, an address, a blood group, etc., and even when it is an accident, it is related with the structure of shoes provided with the information portable function which can perform a check of the person himself/herself reliably.

[Background of the Invention]

[0002]

When an accident occurs outdoors, specification of the victim's individual is required, but the case where the specification is difficult has actually occurred.

When many injured and the dead are simultaneously reported like especially an earthquake and a tidal wave, the check of the identity and specification are very difficult.

Therefore, the check must be easy, if the card etc. which described personal information are always put into a pocket and are carried.

[The outline of a device]

[Problem(s) to be Solved by the Device]

[0003]

However, ordinary persons possess various kinds of clothes fairly according to a season, or liking and the purpose.

It is almost impossible to have the card as which those persons wrote down information in all the clothes, and to always walk around with it.

Since clothes wash frequently, its act of taking out a card at every time and storing after wash is not very troublesomely more practical still, either.

[Means for solving problem]

[0004]

The structure of shoes provided with the information portable function of this design made in order to solve the above conventional problems, It is a back surface of the upper of shoes, and the card stowage container which constituted the loading slot from a waterproof water proof zipper at the back surface of a tongue, a tiptoe hide, and a point core is provided, and an ankle conclusion belt is attached and constituted in some shoes.

[Effect of the Device]

[0005]

Since it came to have described above the structure of shoes provided with the information portable function of this design, it can obtain at least one of the following effects.

The buyer who purchased the shoes of the composition of <1> this design inserts and uses the card etc. in which the matter proving a status was written down for a container with a lid.

Since shoes are always worn, and shoes can be hard to come off and are moreover constituted from a string, rubber, etc. at the time of <2> and going out, it is hard to separate from a victim's body.

<3> At this point, if the pocket of underwear, a coat, and a coat is equipped with social position information, the same effect should be acquired, but as described above, clothes possess many mark actually, and since there is also much number of times of wash, it is difficult to always carry a card.

<4> There is little man in the street who has much various kinds of shoes

according to the season, or liking and the purpose to it like [there is not so many possession of shoes and] clothing, and there are also few opportunities of wash.

<5> If the stowage container with a lid is therefore provided in shoes, social position information is securable only by saving information in the shoes of several points.

<6> The shoes further provided with the information portable function of this design equip some shoes with the ankle conclusion belt. Therefore, for example, even if it is involved in a tidal wave which is expressed as "a shock which was involved in the washing machine" and shoes are able to come off from a foot, shoes cannot secede from an ankle and can hold information reliably. Although neither <7> children, nor a student, an elderly person, etc. have an identification card like a driver's license in many cases, when shoes are not worn outdoors, there are rash. [no] Therefore, even if it should have an accident when having worn shoes provided with the information portable function of this design, individual information can always be carried.

[Brief Description of the Drawings]

[0006]

[Drawing 1]The explanatory view of the working example in shoes.

[Drawing 2]The explanatory view of the working example in sports shoes.

[Drawing 3]An explanatory view when shoes are able to come off.

[Working example]

[0007]

Next, it describes about the working example of the structure of shoes provided with the information portable function of this design.

[0008]

Structure of <1> shoes

If it describes about the structure of common shoes first, shoes are classifiable at the upper 1 and the bottom 2.

That is, it is a general term of the portion after removing the bottom 2 in the upper 1.

And the shoes made into the object of this design can target all the shoes, such as shoes, sports shoes, and high-heeled shoe.

[0009]

<2> stowage containers

It is the bag body constituted from waterproof packaging in the stowage container 3.

The opening of a freely openable/closable loading slot like the water proof zipper 31 is carried out to the part.

The card in which personal information was written down can be inserted from the water proof zipper 31 portion.

Personal information is the name of the person himself/herself, an address, a telephone, a blood group, a previous illness, a relative's name, a school, a company name that have worn shoes.

[0010]

<3> fixing places

The stowage container 3 is installed in some back sides of the upper 1 of shoes by adhesion, tuck, etc.

For example, they are a back surface of the portion which is called the tongue 11 in the case of shoes, a back surface of the tiptoe hide 12 or a back surface of the point core 13, a back surface of the heel, etc.

In sports shoes, if a fine sight is not spoiled in design, the aforementioned stowage container can also be attached to the surface of the upper 1, i.e., the upper surface of the upper 1, a side surface, and a back face.

Also in such a case, the card of personal information will not become dirty or the loading slot of the stowage container 3 will not be damaged, if it protects by the water proof zipper 31.

[0011]

<4> ankle conclusion belt

The ankle conclusion belt 4 is attached to a part of structure of shoes provided with the above-mentioned information portable function.

The ankle conclusion belt 4 consists of a belt with long usual fiber, synthetic fiber, material expanded and contracted, chain, etc., and a string.

By the present invention, the ankle conclusion belt 4 is called also including material only with a strip-like material corded such.

[0012]

Attachment of a <4-1> ankle conclusion belt

When circular "knob" 14 for pulling up when wearing shoes are attached, through the knob 14, the ankle conclusion belt 4 inserts the ankle conclusion belt 4, and is attached to the heel side called a heel cup etc.

Since the thing of the structure where pinch at the end of a tongue in sports shoes, and

14 is attached is also marketed, the ankle conclusion belt 4 is inserted and attached through the knob 14.

A short cylindrical body which lets an elastic band pass is attached to the heel side, the cylindrical body is pinched, as 14, it is made to penetrate there in the case of shoes without the knob 14, such as shoes, and the ankle conclusion belt 4 is attached to it.

It is possible not only when making it penetrate, but to attach by point contact which sews a part on.

As furthermore shown in Fig. 3, the ankle conclusion belt 4 which made the knob penetrate at the time of wear can once be tied with the circumference of the knob 14, and the ankle conclusion belt 4 can also be twisted around an ankle after that.

[0013]

The length of a <4-2> ankle conclusion belt

The length of the ankle conclusion belt 4 carries out the ankle at least 1 round, when shoes are worn, and it is a front part or length where it is possible to connect at rear end.

If the interview catalyst object called Velcro (registered trademark) etc.

is attached to some ankle conclusion belts 4, since it is necessary not to dare to connect, so short a dimension is sufficient.

[0014]

<4-3> fashionability

If this ankle conclusion belt 4 is constituted from the rich belt and string of color, when twisting this around an ankle, it becomes a fashion with which a step is decorated.

If flower conclusion is used around the knob 14 behind shoes even when not twisting around an ankle, commodity value can be raised as shoes ornamented but [so].

When many shoes are put in order in the shop and are furthermore sold, since only the shoes of this design have attached the fashionable ankle conclusion belt 4, a purchaser's attention can be pulled, and they can attain differentiation with similar commodities, and can contribute to sales largely.

[0015]

The display effect of a <4-4> ankle conclusion belt

When going out, naturally shoes are worn, but there are few people who also usually hold tens of pairs of shoes, and they are about at most several pairs of shoes.

If it is the number of shoes to that extent, it is not so difficult to insert the card as which personal information was written down in the stowage container of all the shoes, either.

Furthermore, except sports shoes, since it hardly washes, once it inserts the card, it can be held and kept as it is.

However, even if it had worn these shoes, it will not be helpful, if it should have an accident and neither the police nor the person in charge of fire fighting notices the information inside those shoes.

in that case -- if it is shoes of this design -- the knob 14 -- being certain -- it is, and it can creep and the ankle conclusion belt 4 is attached to that position.

Therefore, I can have you able to recognize it as personal information being kept if it is a certain shoes which the ankle conclusion belt 4 attached when the function of the shoes of this design and the effect were known widely, and an identity can be checked promptly.

[0016]

When <4-5> shoes are able to come off (Fig.4)

As described above, like the shock of a tidal wave is expressed saying, "It was involved in the washing machine", it is intense, and clothes etc. may be stripped off.

The automobile of reckless driving, therefore also when it encounters, shoes bound off with the shock of a collision.

Therefore, if it is common shoes, a possibility that the worn shoes will be wrested away will also be assumed sufficiently.

However, in the case of shoes provided with the information portable function of this design, shoes are firmly attached with the ankle conclusion belt 4 at the ankle.

Therefore, even if there is a fairly big shock, since it is larger than an ankle, shoes cannot

separate from a foot, and the length from the heel to a tiptoe can perform the original function to show this humanity news.

[Explanations of letters or numerals]

[0017]

1: Upper

11: Tongue

12: Tiptoe hide

13: Point core

2: Bottom

3: Stowage container

31: Water proof zipper

4: Ankle conclusion belt

Drawing1

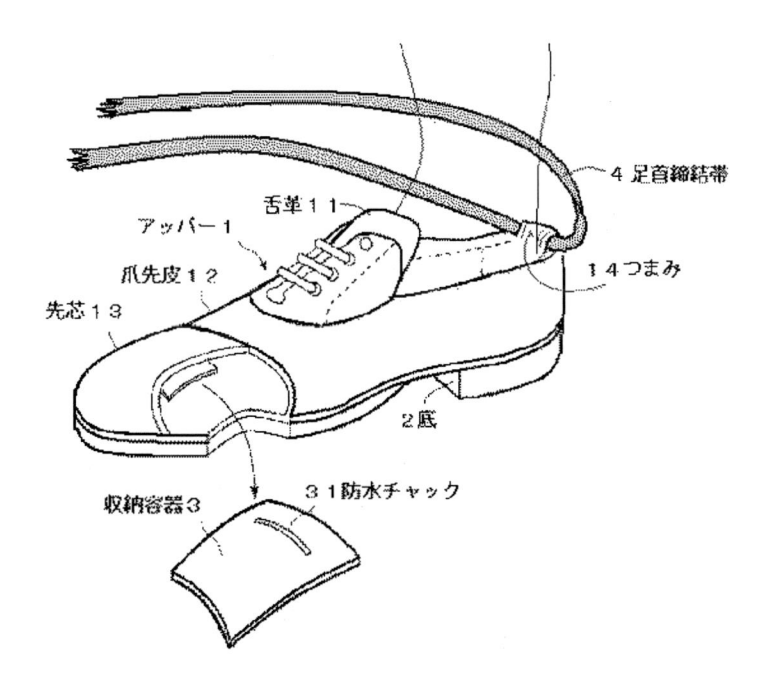

Drawing2

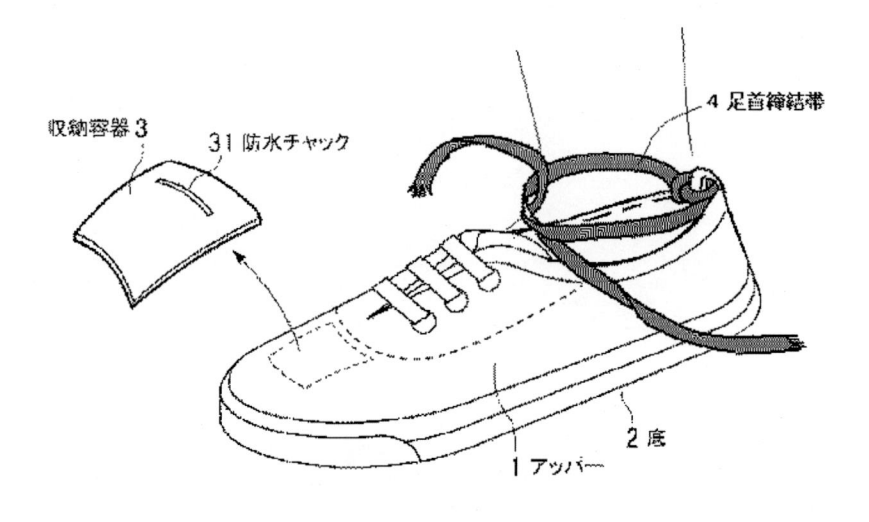

収納容器 3　　31 防水チャック　　4 足首締結帯

1 アッパー　　2 底

平常時に靴を履いている状態の説明図。
（靴の内部には個人情報カードが入っている）

The personal information card is contained in the inside of shoes.

44

Drawing3

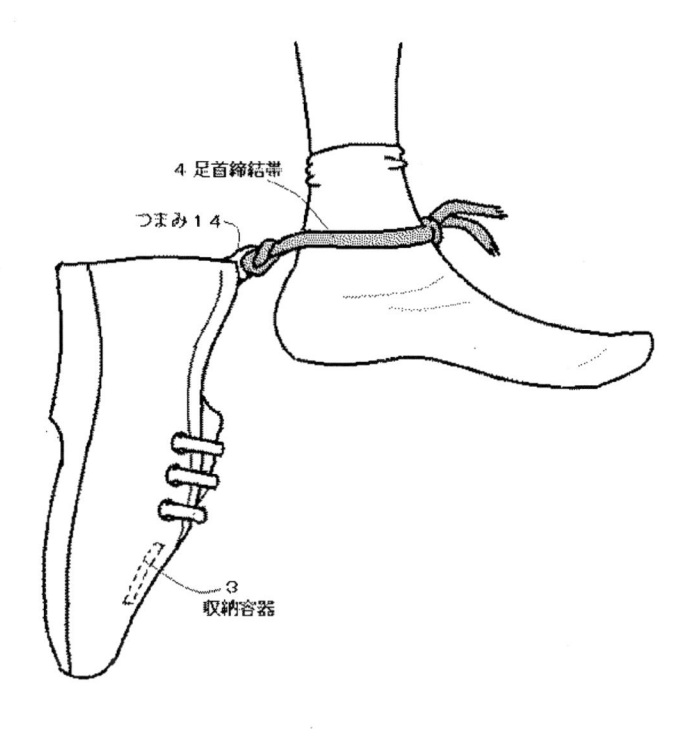

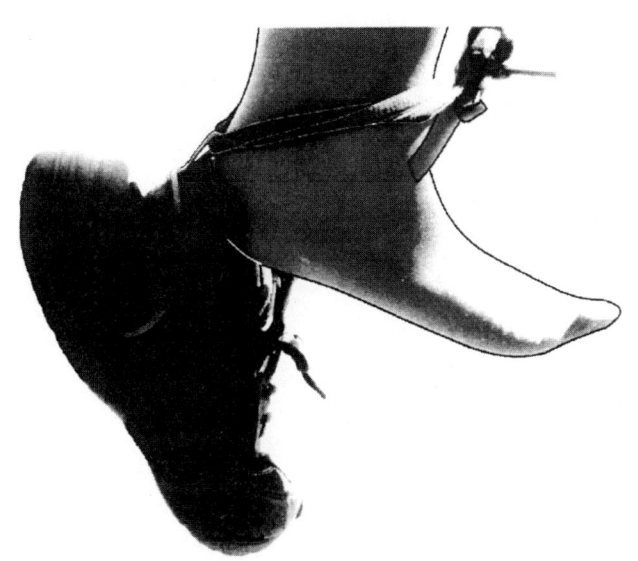

非常時に靴が脱げた状態の説明図。
（靴の内部の情報カードが、足から離れない）

The state in which shoes were able to come off.

The personal information card inside shoes does not separate from shoes.

あとがき

「ナゼ？」

旅から帰る自宅周辺に来ると、気持ちが"ホッ"とする気持ちはナゼ？

旅行から自宅に帰る安心する気持ち、それは逆に考えると、事故、もしくは事件にあう事が気持ちの中に入っていると思う。だから自宅に帰ると安心する気持ちが沸いてくる。

現在、私達が皆が旅行、または会社に行く途中、何らかの事故・事件・病気などに身元が早く解るのは靴。

それによって、家族・親戚に連絡が出来る特許の靴が誰も、自然な形で利用される事が大きな災害の時に大きな役割を果たす。

著者　斉藤　通
（さいとう　とおる）

天災・人災 被害者の身元確認情報靴　情報携帯靴 ワカール

定価（本体1,000円＋税）

２０１３年（平成２５年）９月１２日発行

No. ST-013

発行所　発明開発連合会®

東京都渋谷区渋谷 2-2-13

電話 03-3498-0751㈹

発行人　ましば寿一

著作権企画　発明開発連合会

Printed in Japan

著者　斉藤　通 ©

本書の一部または全部を無断で複写、複製、転載、データーファイル化することを禁じています。

It forbids a copy, a duplicate, reproduction, and forming a data file for some or all of this book without notice.